LIFE IN THE LIBERATED ZONE

DAVID LURIE

TEXT BY
RIAN MALAN

Cornerhouse Publications, Manchester, England
William Waterman Publications, Rivonia, South Africa

LIFE IN THE LIBERATED ZONE

DAVID LURIE was born in 1951 in Cape Town. He studied politics, philosophy and economics before leaving for London in 1980 to study further and work there. In 1988 he began work as a professional photographer. His photographs have been published and exhibited to widespread critical acclaim in Europe, America, Australia and South Africa.

Life in the Liberated Zone received several Ilford Profoto Awards (South Africa) in 1992 as well as an American *Pictures of the Year* 'Award of Excellence for Magazine Picture Story' (1992).

He is completing his second book, on South Africa's so-called 'Black Middle Class', provisionally titled *Now it's Our Turn!*, with text by Eskia Mphahlele.

RIAN MALAN, a member of one of Afrikanerdom's oldest families, was born in Vereeniging in 1954. Educated at Woodmead High, he joined *The Star* and became their crime reporter. In 1977 he left the country for America. He returned from exile in 1986 and resumed work as a crime reporter.

He went on to write the internationally acclaimed *My Traitor's Heart*. He now writes for leading magazines around the globe and is currently researching a book on tribalism.

Rian Malan: my story

I first saw the landscape where these photographs were taken in the summer of 1965, through the windows of my father's car. My father was a bureaucrat in a big company. In 1965, he was transferred from Johannesburg to Cape Town, where we lived for a year in a flat, in a suburb called Rondebosch where it rained a lot. On Sunday afternoons, we would often go on drives to look at the scenery. Sometimes we went south, through the oak forests to Cape Point. And sometimes we went the other way — past Groote Schuur Hospital, where Dr. Chris Barnard was planning the world's first heart transplant; past Athlone power station, past factories and tank farms and chemical lagoons, and out onto the windswept Cape Flats.

In 1965, the flats were empty and desolate. There were no farms, because the soil was beach sand. There were no houses, because the flats turned into a swamp during the winter rains. There were no tall trees, because the region lay open to howling southeasters that snapped anything taller than ten feet. All you saw from the car window was a low wall of wind-stunted scrub, tangled and impenetrable, and a sky full of seagulls and jet planes, coming in to land at DF Malan airport. It took about thirty minutes to cross the flats, and I was always glad to reach the other side, where there was a pretty town called Stellenbosch, and better things to look at — lush green vineyards, gracious homesteads with Dutch gables, groves of oak and chestnut trees.

I loved the Cape when I was ten. It was like another country, another continent, far too green and wet and gentle to be part of Africa. Also, there were virtually no black African people. You hardly ever saw a black face on the street. This was a bit perplexing to a ten-year-old from Johannesburg, because Jo'burg was full of blacks. We'd had two in our service, a gardenboy named Johannes and a maid named Lena Tshabalala. In Cape Town, all we had was Annie, a mixed-race coloured woman who came to clean our flat once a week. Annie was very different from the blacks I knew. She spoke Afrikaans, just like me, even had a Boer last name. She was very self-confident, very savvy, and completely disinclined to take shit from whites, not that my mother would have dreamed. Also, her services were very expensive. That is why she only came once a week. That is also why I had to make my own bed in the morning, and mop the kitchen floor on Saturdays.

These were novel experiences for a ten-year-old white boy, and made possible, in a round-about way, by apartheid. Our Calvinist leaders had determined, following close study of the Old Testament, that there were actually four different kinds of people in South Africa. First, there were whites like me and my family. We were clearly in charge of things. Then there were blacks, who were very poor, spoke broken English and usually lived in whites' backyards. In Natal, there were a million or so Indians, and finally, there were coloureds like Annie, who were the descendants of Boers like ourselves and the Cape's original Khoisan inhabitants — our own flesh and blood, in other words, but spurned by apartheid after 1948, stripped of their vote, banished from white schools and suburbs, forbidden to marry white people.

As a supposedly separate and distinct nation, the coloureds were entitled to their own "place in the sun" under apartheid, and this place was the Western Cape, a Coloured Labour Preference Area in law. It was virtually impossible for blacks to get passes to live in Cape Town, and nobody was allowed to employ a black person unless it could be proved that no coloured person was available. There were only seventy thousand blacks in the entire Western Province, and they were all "temporary sojourners", in apartheid parlance. If they lost their jobs or retired, they were required to go "home" — over the mountains, across the arid Karoo, over more mountains still, to places called Transkei or Ciskei.

Transkei and Ciskei were the ancestral territories of the Xhosa people, who fought nine bloody wars against the Boers and British in the 19th Century. In the end, their military power was broken, but they somehow managed to hold onto about two-thirds of their ancestral land, a swathe of territory about the size of Denmark. It was good land, for the most part, but there was too little of it, and by 1965 it was overcrowded and ecologically damaged, riven with erosion scars, the plains grazed flat as a billiard table by marauding goats and cattle. The land could no longer feed the people, so men had to go away to the cities and work for whites. They went to the mines near Johannesburg, or to the factories in East London or Port Elizabeth, but they could not come to Cape Town, because Cape Town was for whites and coloureds only. And that's how it was, in South Africa, in 1965.

We drove through the Transkei a few times when I was a child, on our way to this seaside resort or that. There were no fences in Transkei, so goats and cattle strayed into the roads. The hills were dotted about with mud houses, and there were ragged peasants on the roadside, dressed in rags and tatters. My mother always encouraged us to wave at them, and they always waved back. And then we drove on, sparing little thought for those left eating dust behind us.

The story of Patience

Her name is Nomazizi Mgwadlega, but she has an alias, a Christian name, so we can also call her Patience. She was born in the Transkei in 1949, in a mud hut near Mount Fletcher, daughter of a man who toiled in Johannesburg's gold mines, and a mother given to fervent singing and praying. She spent seven years at a mission school, learning the rudiments. At the age of 16, she fell in love with a dashing young man named Cyril Ntamo. They slept together once or twice, and then he went off to Johannesburg to work in the mines, to earn money to buy cattle to pay Patience's bride-price. When he came back, a surprise was waiting: twin sons, Robert and James. Cyril was ecstatic. He gave Patience's father the first of the agreed-upon eight cattle and took her and the babies to live in the kraal of his family, across the valley. One month later, he returned to the mines, leaving Patience and the family behind. That was the law; dependants were forbidden to join apartheid's labour units in the cities.

At the outset, Cyril was a good husband. He sent money. He wrote letters. He came home every Christmas, reducing his sons to tears of fright and bewilderment, because a year is a long time in the life of a child, and the man who returned had become a stranger. Two more children were born, and Patience was content with her life, in spite of the long separations. And then, in 1976, something happened. Cyril's letters tailed off, and then ceased entirely. He failed to come home for Christmas. He stopped putting money into the Post Office. Patience was forced to rely on the charity of her in-laws, who had very little to spare. Her anxiety changed into anger, and then despair. In the end, she boarded a bus and went to gold mines near Johannesburg to look for Cyril, but she found no trace. Perhaps he had started a new life with another woman. Perhaps he had been killed in a knife fight, and been buried in a pauper's grave. Perhaps he had gone to jail. She never knew. All she knew for certain was that her husband had vanished, leaving her with four young children to feed.

But how? Life was very hard in a Transkei village. You could not farm because the land was finished. There were no jobs in hundreds of miles. You had to walk to fetch water and carry it home on your head. You had to walk to fetch wood, sometimes all day. If your child fell sick it would most likely die. There was no money, little food. There was no hope, even. There was only Cape Town. Patience had heard that there were jobs to be had in Cape Town, menial jobs in kitchens and factories, drudgeries to which whites would never stoop and which coloured people had come to regard as beneath their dignity. She knew that it was forbidden for a black person to go there without a pass, but what could she do? The children were crying. There was nothing to lose. So she packed her bags, padlocked the door and boarded a bus bound for Cape Town.

The journey from Transkei to Cape Town was not so very different from the *via dolorosa* between Mexico City and San Diego, or Istanbul and Berne — hopelessness at one end, hope at the other, and in between, roadblocks manned by policemen whose task was to identify illicit migrants and turn them back. Three times, the bus was stopped by police and searched for illegals. Three times, Patience talked herself out of trouble by claiming to be en route to Cape Town to visit a relative. In a way, this was true. She had a piece of paper on which her father-in-law had scribbled the name of a distant clanswoman who was working legally in Cape Town. The policemen studied it sceptically and then shrugged, okay, you can pass; just make sure you go home again, or we'll nail you. And so she got through. The bus dropped her at the station. It was cold. The wind was blowing. Her youngest child was with her, a toddler named Elizabeth. She took her daughter by the hand and turned her face towards the Cape Flats, where I found her sixteen years later.

Her shack stood not far from the road we had once driven on those childhood Sunday drives, the road that led past Groote Schuur, past the factories and toxic lagoons, and out onto the windswept flats. The sky was still full of scudding clouds and jets and seagulls, but the earth beneath was now carpeted with wooden shacks and rusting tin shanties, hundreds of thousands of them, extending in all directions, as far as the eye could see. I arrived towards sunset on a winter's evening. A gale was whipping across the shacklands, carrying the salt tang of sea, the acrid whiff of cooking fires, the lowing of cattle, and the crack of plastic sheeting whipping in the wind. Patience was standing in the doorway of her wood and iron shack, staring apprehensively at the sky. "I hate this blowing," she said. It reminded her of bad things, of death and fire, of nights spent shivering in the open. She's had a hard time, these past sixteen years. She puts a kettle on the Primus stove. She invites you to sit down, and begins to talk about it.

The story of apartheid, and Patience's part in its downfall

Once upon a time, on the fringes of Cape Town's industrial zone, there was a place with no name where two roads met. The land surrounding the intersection belonged to the government, and for years, it just lay there, clad in dense bush, awaiting the onward march of industrialisation. Nobody quite remembers when the first shacks appeared, but it was long ago, perhaps as early as the 1940s. The first squatters were legal residents of the Cape, young couples or poor people who could not find accommodation in the overcrowded townships. In the 1970s, numbers grew, and a new life-form appeared; desperately poor migrants from the homelands. By the time Patience arrived, the place had a name, Crossroads, and a population of perhaps 20,000 passless illegal migrants, living in primitive structures in the bush and sand dunes.

On the day she arrived, Patience borrowed an axe and bought a few rolls of black plastic garbage bags. She cut poles in a thicket, sunk them into the sand and lashed the ends together. She sheathed this frame with garbage bags, to make it rainproof, and lined the interior with cardboard, for warmth and insulation. The floor was sand. It was a sorry, flimsy little structure, but it would have to do. Patience moved her belongings inside and set off for the white suburbs, in search of a job as a maid.

In this respect, she was lucky. Mrs. Berman hired her to clean three times a week, and Mrs. Van Staden needed a washerwoman on an occasional basis. The combined money was enough to live on, and there was even a little extra, to support the children she'd left at home. It was a satisfactory beginning, in Patience's estimation, but the good times soon ended. In the eyes of apartheid ideologues, Patience and her neighbours constituted a challenge to the existing order, and a threat to their dream of a country starkly divided into various ethnic territories. The mere presence of black squatters on the outskirts of Cape Town was disturbing, a sign that something had gone awry with the grand apartheid masterplan. Blacks were supposed to be draining away into the tribal homelands, not flooding into Cape Town, squatting in a whites-only "group area" and taking jobs reserved by law for coloured people. Crossroads threatened to make a mockery of apartheid, which is why it had to be destroyed, by almost any means necessary.

Patience remembers the first raid quite clearly, if only because the experience was novel. They came at dawn in a convoy of trucks, a troop of officials from the Department of Bantu Administration, backed up by a squad of policemen. She heard shouts in the distance, the barking of dogs. She looked outside, and saw people running away through the greyness, so she scooped up a child and ran away too. The authorities moved through the camp, breaking shacks down, confiscating anything that could be used for building material. They loaded these on the back of the trucks, along with a handful of captives, and drove away. Patience crept out of hiding. Her blankets, food and pots and pans were lying in the open, in a tangled wreckage of torn plastic and broken poles. What could she do? She was black. She had no rights, no avenues of appeal. She just took up her axe and cut more poles. By nightfall, she'd completed a new dwelling. A week later, it was destroyed in another raid, and destroyed again soon after.

After a while, the raids began to blur together. Patience had no idea how many times her home was destroyed, rebuilt and destroyed again; how many friends and neighbours were arrested and deported to the homelands in handcuffs. All she remembers is that the struggle continued for years, with occasional changes in tactics. Sometimes the raids were suspended while authorities tried to persuade the squatters to go home voluntarily. When the talks stalemated, as they always did, the raids would resume. After a while, the Boers realised that daylight raids were a waste of time, and began to invade at night, hoping to catch the illegals in bed. In response, the squatters took to sleeping in the bush when a raid was imminent.

It was a miserable life, but Patience bore it without complaint. In some ways, she even considered herself fortunate. She was never once arrested and deported. She and her daughter spent many nights shivering under the stars in the wet and cold, but they never contracted tuberculosis or pneumonia. Her madams lent her money, replaced clothes and blankets lost in the ceaseless raids. In the early eighties, she met and married Zephaniah Mpeta, an older man who held a good job in a plastics factory. Zeph lived in an older part of the Crossroads, in a community so long established that the authorities had ceased trying to uproot it. His shack had a corrugated iron roof, four rooms and floral wallpaper, even a TV set powered by a car battery. It was almost a real home, its virtual permanence a sure sign that persecution was flagging.

By 1983, the white authorities were beginning to despair. Every time they tore a shack down, another two would spring up in its place. Every illegal arrested and deported came back again, usually with several family members in tow. Worse yet, every raid was turning into a media event. Liberals from the white suburbs had taken to forming human chains around the squatters. Reporters captured damning footage of shacks crushed under bulldozers, mothers dragged off in handcuffs, their children left shivering and crying in the rain. These images caused an international outcry, undermining State President Botha's attempts to present a reformist face to the world. In desperation, Botha opted for a policy of controlled organisation: illegals already established in the Cape Flats squatter camps would be forgiven their transgressions against apartheid and granted the right of legal residence. As word of this concession spread, the inflow from the homelands redoubled, and apartheid's defences began to crumble. In Easter 1986, the government capitulated, and the hated pass laws were scrapped.

After that, it was as if a distant dam had broken, allowing a mass of desperate and hopeful humanity to come flooding over the mountains and spread out across the flats. They came at the rate of eighty, ninety families a day, and built homes with their bare hands, using wooden poles, tin sheeting, bits and pieces of trash rescued from landfills and plastic garbage bags to keep out the rain. Within two years, the sand dunes had vanished under an enormous sea of shacks and shanties, as densely packed as a mediaeval city, and populated by fantastic characters — bootleggers, gangsters, prophets, Rastafarians, gun dealers and marijuana czars, plus almost a million ordinary working people.

Crossroads was a violent, chaotic place, wracked by appalling social problems, but its grimness was by no means relentless. Amazing sights awaited the visitor brave enough to wander its narrow alleys — tiny shops and bars; open air restaurants where chefs roast sheep heads with blowtorches; immaculate cottages with rose gardens and lace curtains in the window; teenage Xhosas daubed white from head to toe, undergoing ancient tribal initiation rites under the flight path of incoming Boeings. Schools and clinics came into being. The government built tarred roads and installed high-mast lighting, in a largely vain attempt to make the streets safe at night. Patience watched these changes with a deep satisfaction, because they were the fruits of a great victory for her people, and a sure sign that white power had weakened to the point of near collapse. And yet, and yet, and yet: Patience would not claim that this was a happy ending. The pass laws were dead, and persecution had ceased, but the cycle of sorrows continued.

Life in the Liberated Zone

It is ten at night, and Patience is brewing yet another cup of tea. The shack is quiet, the children and grandchildren sleeping. Outside, the wind is howling, and there are ominous sounds in the distance, faint cries, and what could be the sound of gunfire. Patience walks to the window, and peers out into the menacing darkness. She is ill at ease.

"It is very dangerous to live in these houses," she says. "There are many gangsters in this place, gangsters and rapists. When it is night, I become worried, and when there is wind, I become worried, because of the danger of fire. If the fire can come in this house, how will I save the children? That way, I do not know how to sleep. I move from one window to another. In the middle of the night you can see me awake, and you would think that I am a witch. When there is heavy wind I do not sleep at all, because I want to be able to get out if there is a fire. This thing is making me to have a sore heart."

And what is "this thing?" It's no longer apartheid, in the sense that apartheid is associated with white domination, with guns and Alsatians and the depredations of heavy-handed policemen. The white authorities lost control of the squatter camps years ago, during the great anti-apartheid uprising of the mid-eighties. Crossroads and its satellites were virtually unpoliceable, their alleys too narrow to allow passage of armoured cars, their comrades too aggressive to be contained. As white rule weakened, power passed into the hands of semi-feudal warlords and revolutionary gangs who fought bloody battles for political control. Ordinary people like Patience and Zephaniah tended to get caught in the crossfire.

The first burning was in 1985, in the time of strikes, stayaways and street warfare between the comrades and the security forces. At the time, Patience and Zephaniah were living under the rule of a strongman named Yamile, to whom they were paying a monthly tribute of R10, in return for which they received protection from enemies and security of tenure. Yamile was allied to the comrades of the African National Congress, which meant that Patience and her family were also aligned to the ANC, and thus at war with Nxobongwana and Nongwe, who were deemed to be enemies of "The People".

Patience had little interest in or understanding of this system of alliances and politics. She was a member of no movement, and her politics were the politics of security and food. But she lived in Yamile's territory, and that was enough to make her a target in August, 1985, when the combined army of Nxobongwana and Nongwe descended on Crossroads, armed with spears and guns and pangas, and laid the place to waste, while the police looked on and did nothing. Almost sixty people were killed, another forty thousand rendered homeless. Patience and her husband lost everything save their children and their precious TV set.

They lived in a tent for a while, and then moved to Black City, which fell under the indirect control of the strongman Nongwe, once an enemy of The People, but now allied to the ANC, and at war with his old sidekick Nxobongwana, now fighting under the colours of the ultra-leftist Pan-Africanists. When darkness fell there was gun-fire in the distance, and fire in the sky. One night, a strong wind was blowing, and the fire burned out of control, swept up the hill and engulfed the house where Patience and her family stayed. They rebuilt it, only to lose everything yet again, in an accidental fire, and again, in the bloody "Taxi War" of 1991, in which a group of transport operators allied to the ANC took on a rival taxi association with nebulous links to black policemen. Patience and Zeph salvaged a few charred remnants from the ashes and carried them here, to this place with no name, where they started again from scratch.

Thus far, there has been no trouble in this area, but you never know, you never know. Zeph grew so weary of life in a low-intensity war zone that he has returned to his ancestral village in the homeland, where he lives on an old-age pension. Patience would dearly love to follow him, but she can't, not with a child in high school and another, her brightest hope, studying to be a teacher at the University of the Western Cape. "I am no longer happy here," she says. "Transkei is my home, and I love it there. One day I will go." Meanwhile, she follows the twists and turns of South Africa's great political drama on her battery-powered TV set, and dreams of a life a little easier than the wretched existence she has led thus far.

What do you want? I ask her. She thinks about that for a while, then says, "I want good jobs for my children, and good police. I want a real house, with real bricks, that cannot be set on fire or shot through. And water taps. And a water toilet. And burglar bards, so the criminals can't come in." These are painfully modest dreams, but even so, Patience sometimes doubts that they will come true. The white government has been promising her a real house for years. Once, there was even a rumour that they had set aside R6 million to build houses for her and her neighbours, but nothing happened, and she suspects that the money was stolen. Now Nelson Mandela is poised to come to power, and he has made many promises, too. Patience devoutly prays for their fulfilment, but she has ceased to place much store in the promises of politicians. "Everybody is expecting, expecting, expecting," she says. "Free houses! Free cars! Free clothes! I will believe it if I can see it." She sighs, and gestures toward her humble surrounds. "Meanwhile," she says, "we are waiting. We will see. But sometimes, I fear we will be living in these shacks forever."

My story, its ending

January 1, 1994. I am in my car, heading eastwards through the white suburbs of Cape Town, on the route I travelled with my father on long-ago Sunday afternoons. Almost twenty years have passed, but these white suburbs are little changed, save for the models of cars in driveways, and the height of the garden trees. In fact, the sweet white life is in some ways sweeter than ever. The Afrikaners have reformed themselves out of power. The ANC's revolutionaries have traded their AK-47s for navy blue suits and boardroom seats. Sanctions are dead, the stock market is booming, and the skies are dark with potential investors. In just 117 days, South Africa will hold its first free elections, and Nelson Mandela will come to power, ushering in what the newspapers herald as an era of peace, justice and prosperity for all. It is, in short, a perfect day.

And then the freeway swoops down the mountainside and shoots off across the flats, past Athlone power station, past factories and tank farms and chemical lagoons, and out onto the windswept Cape Flats, where rusting shacks stand shoulder-to-shoulder in an endless swamp of misery. There are heavily armed soldiers on the bridges, and sentinels in towers on the freeway verge. They are there to protect people like me from those in the shacklands, and, in truth, I am grateful. I would never dream of visiting Patience on a public holiday, when people have been drinking, and fuses are short, and white skin might turn out to be a death warrant. I wish it were otherwise, but that's how it is, in the dying days of the *ancien régime*.

In a way, David Lurie's images raise the single most important question facing the "New South Africa", the question on which our collective fate may depend. Close on a million people live out there in shackland, in a vast cesspool of poverty and misery. Another million squatters have settled in the veld around Johannesburg, and there are two million in the bush surrounding Durban. There are squatters on the edge of every white town, squatters in parks, squatters on landfills — eight million squatters, all told; about a fifth of South Africa's population, trapped in appalling poverty. The murder rate in their communities is among the world's highest, as is the incidence of tuberculosis. A majority of squatters have no jobs, no food for their families. Some don't even have water. Children are malnourished, riddled with disease. Crime is rampant, and AIDS threatens a decimatory plague. Life in the camps is wretched, and worsening daily.

What becomes of these people now, after the elections? A change of government will not in itself transform the lives of the poorest blacks. You can't eat a vote, or use it to clothe your children. The squatters need jobs, schools, clinics, sewers, electricity and housing. Who will provide such things? Stock market euphoria notwithstanding, South Africa's economy is in poor shape, its population exploding, its capital fleeing to safer climes. It could be argued that the previous government lacked the compassion to help the squatters, but does this democratic black one lack the means, and what then? Another revolution? A descent into looting and banditry, with squatters seizing at gunpoint what was not delivered by the ballot? There is scant hope of lasting peace for anyone in South Africa, black or white, until there is hope and prosperity in the squatter camps, and that will require a virtual miracle. But stranger things have happened.

And Patience is waiting.

Overlooking Khayelitsha, Site B. Early morning, January 1991. Khayelitsha means 'our home'

Nolungile Station, early morning. Boarding trains to go to work

Khayelitsha, early morning. Waiting at the road side to see if any employers are recruiting workers

Recently arrived from Ciskei homeland. Late afternoon, 'Lusaka'. Officially called Miller's Camp, squatters renamed the settlement 'Lusaka' (capital of Zambia) where the ANC's headquarters were based before it was unbanned in 1990

Early morning, New Crossroads

'Lusaka', late afternoon. Cooking offal (entrails) at the roadside

'Lusaka'

Early morning, New Crossroads

Taxi drivers waiting for early morning clientele. A ruthless war is being fought between rival taxi associations over control of lucrative routes

Early morning outside the men's hostel in Langa, Cape Flats. Selling home-brewed beer

Midday, Khayelitsha, Site B

Outside the men's hostel, Guguletu. Cooking sheep's heads which sell for R5 each (about £1.00)

Outside the men's hostel, Guguletu. Cooking sheep's heads

Late afternoon, outside Guguletu men's hostel

After work, 'Black City' squatter camp

“Ek het vok-all!” (“I’ve got fuck-all!”). Late afternoon outside a shebeen (speakeasy) in ‘Lusaka’

Clement, a social worker employed by Operation Hunger, is told about some of the problems in 'Lusaka'

Nyanga township, after work

Brown's Farm squatter camp

Late afternoon, 'Lusaka'

Interior of shebeen in 'Lusaka'

Girl looks after younger brother while their parents are away at work

Eating an orange. Late afternoon, 'Lusaka'

Inside the men's hostel, Guguletu. These 'single-sex' hostels were built by the government in the 1950s to house migrant workers from the homelands

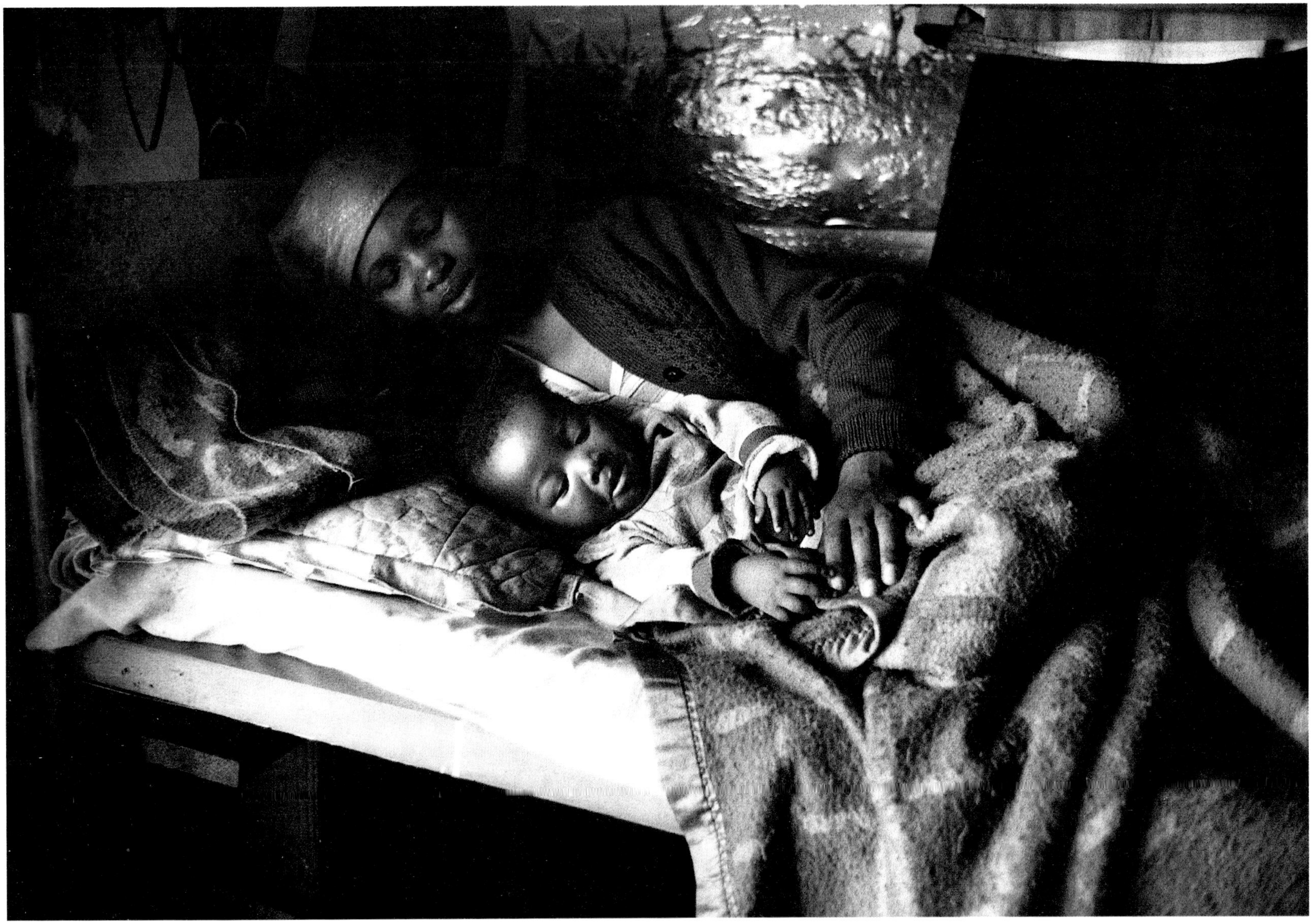

Inside Guguletu men's hostel. The hostels had six beds to a room, each bed is now used to accommodate a whole family

Inside Guguletu men's hostel

Men's hostel, Langa

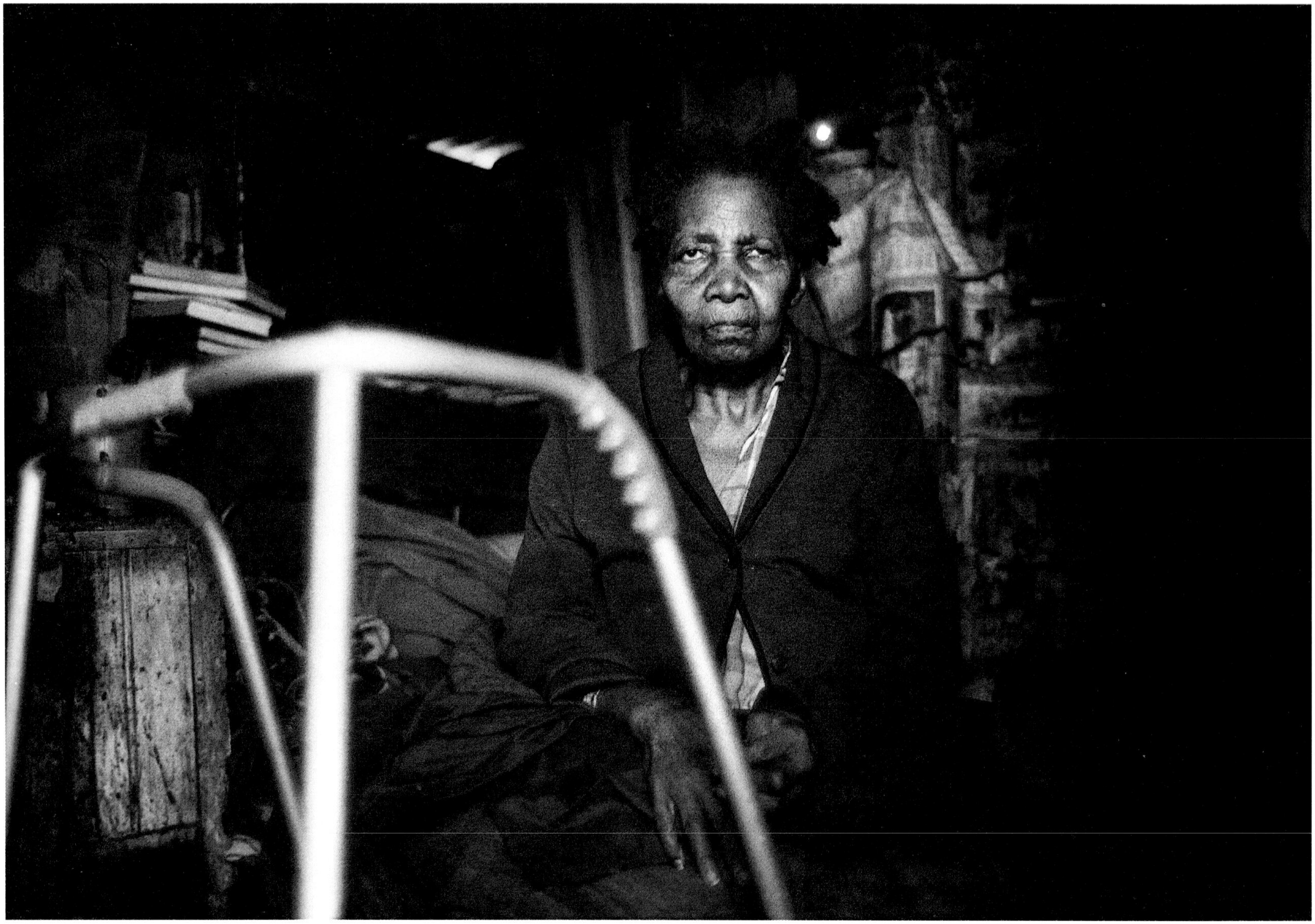

Disabled woman living alone in Nyanga Extension

Interior, Guguletu men's hostel

Interior of shack,
Mpheta Square squatter camp

Interior, Guguletu men's hostel

Children's crèche, Khayelitsha.

Children's crèche, Khayelitsha

Midday, interior of shack, Brown's Farm squatter camp

Midday, another interior of shack, Brown's Farm squatter camp

Interior, Guguletu men's hostel

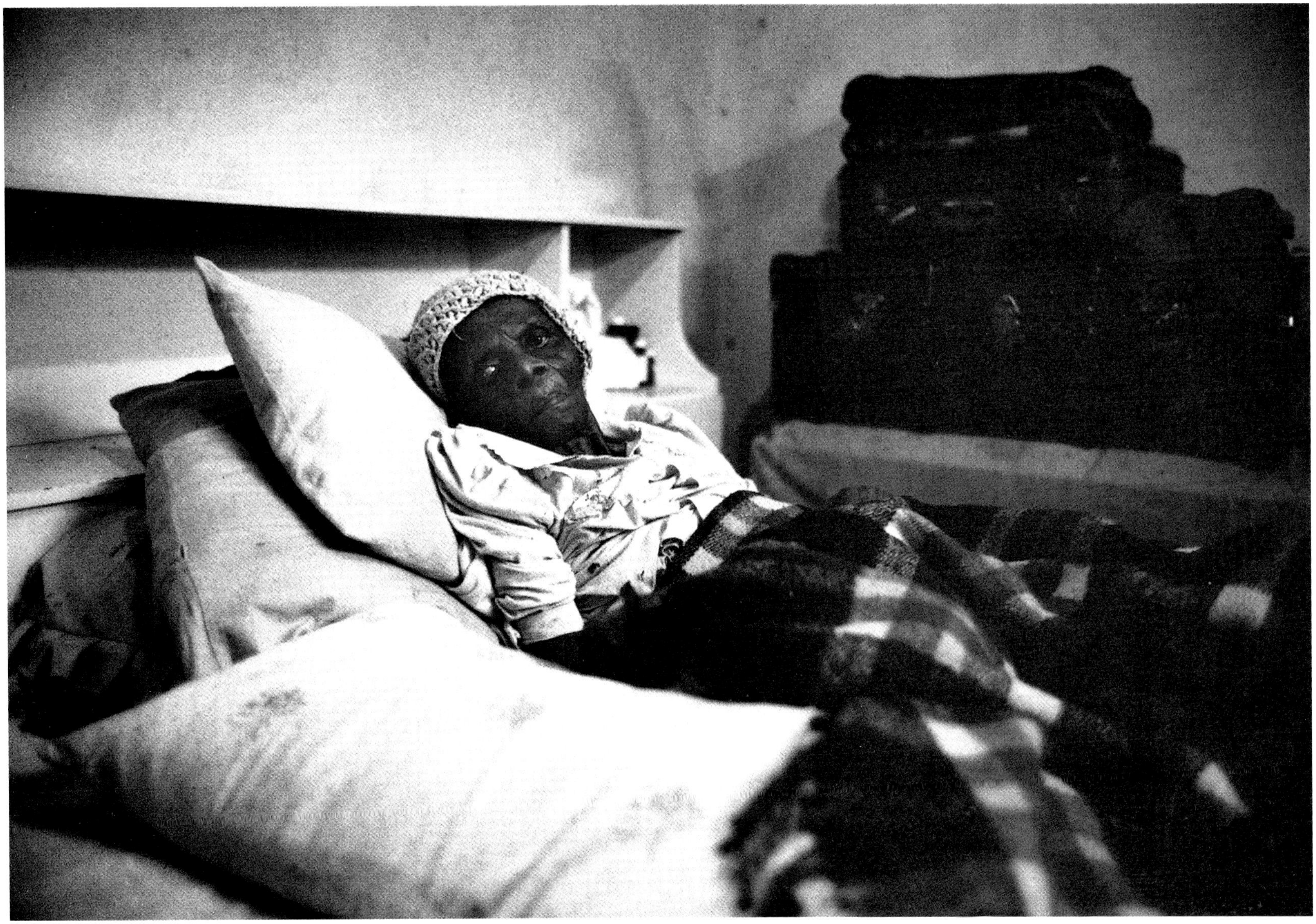

Interior, Guguletu township

Smoking a ‘white pipe’ (dope and mandrax) at a shebeen in Black City squatter camp

Inside men's hostel in Langa, after work

Brewing beer, men's hostel in Langa

Saturday afternoon at a shebeen in Khayelitsha

Late afternoon,'Lusaka' squatter camp

Children's crèche, Khayelitsha

Measles Inoculation Clinic, Crossroads

Interior, Guguletu township

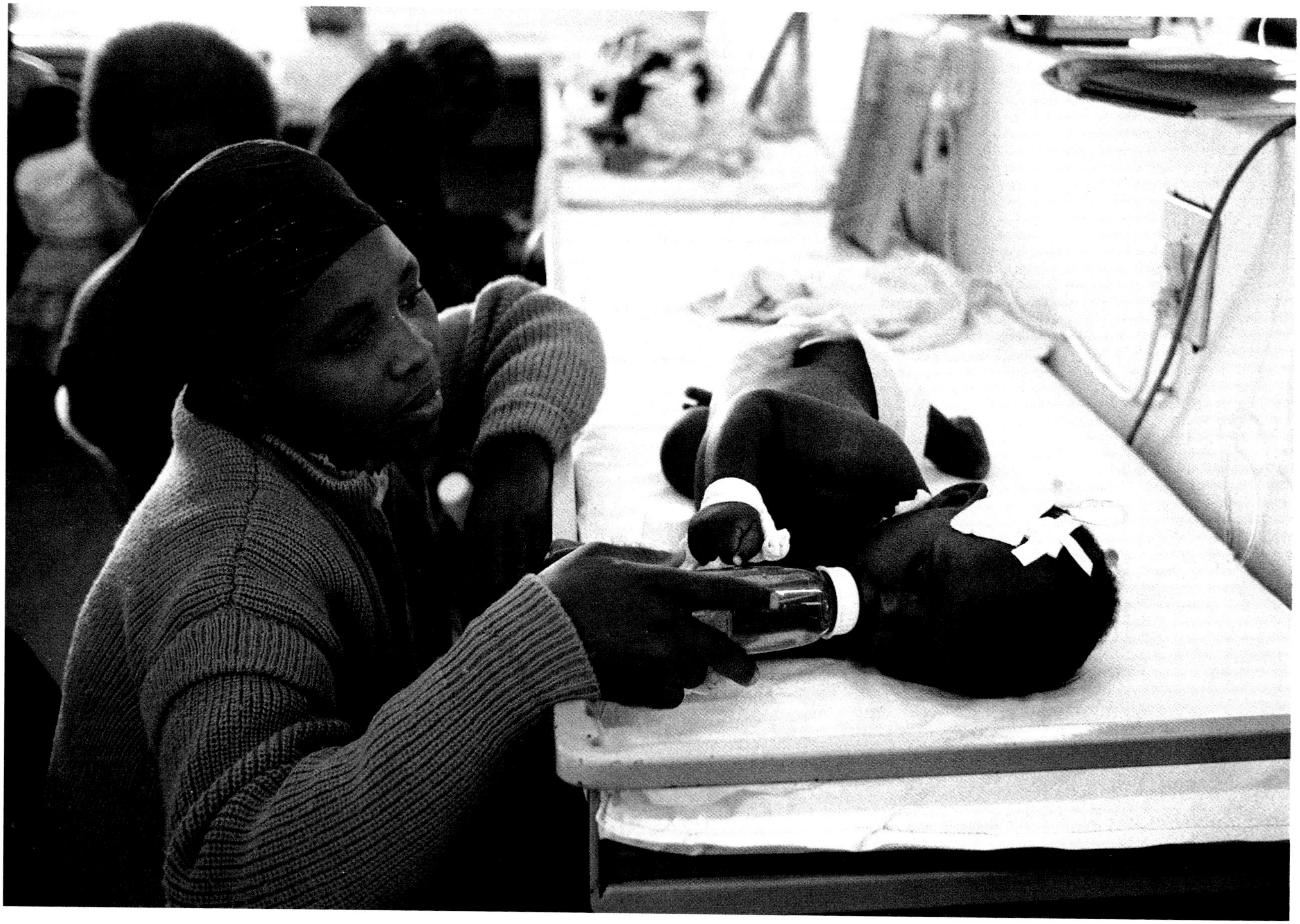

'Drip room', Red Cross Children's Hospital; for rehydrating infants with chronic diarrhoea (caused by infection arising from crowded living conditions with no access to indoor tap water)

Nutrition centre, Black City squatter camp

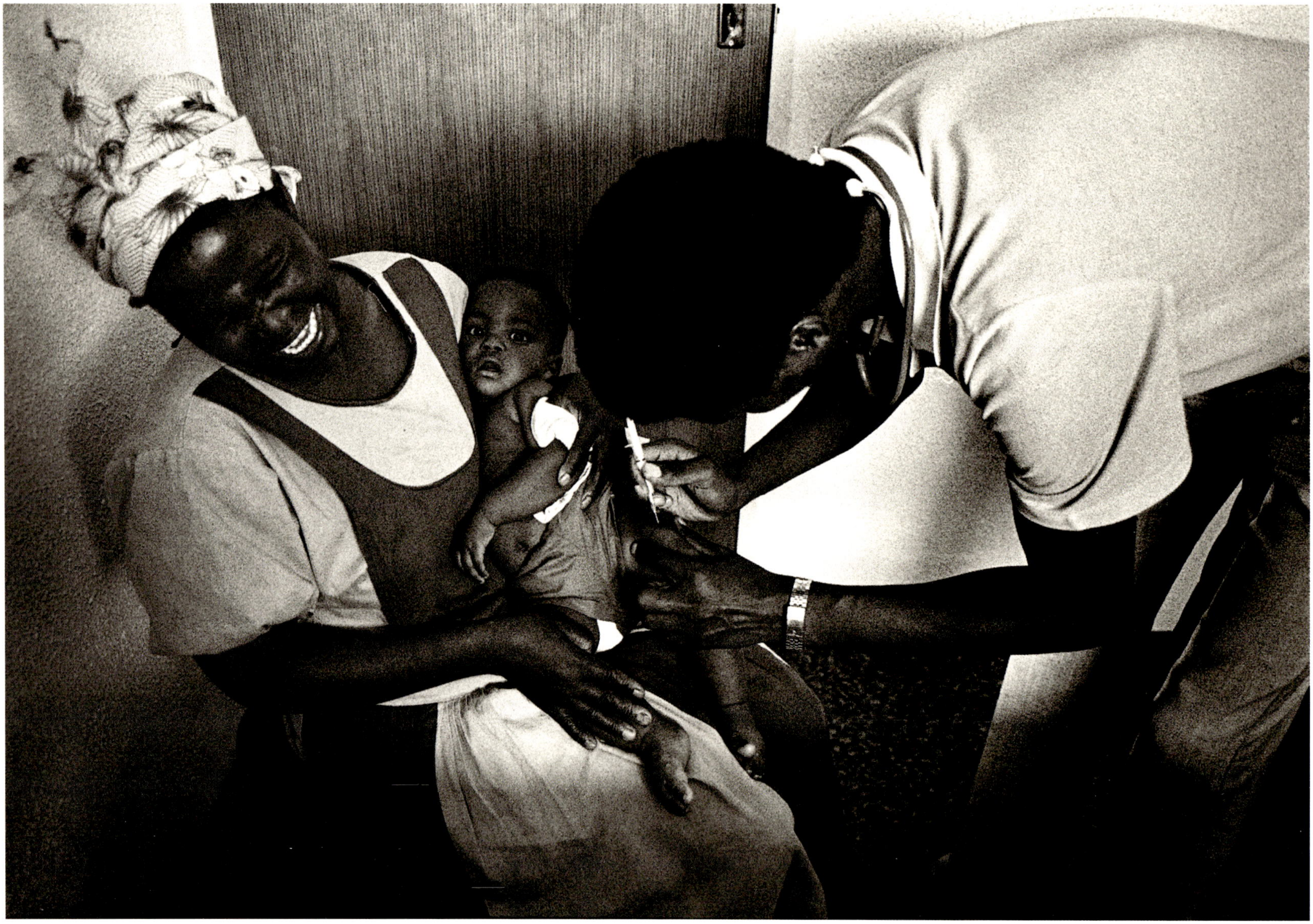

Measles inoculation, Crossroads

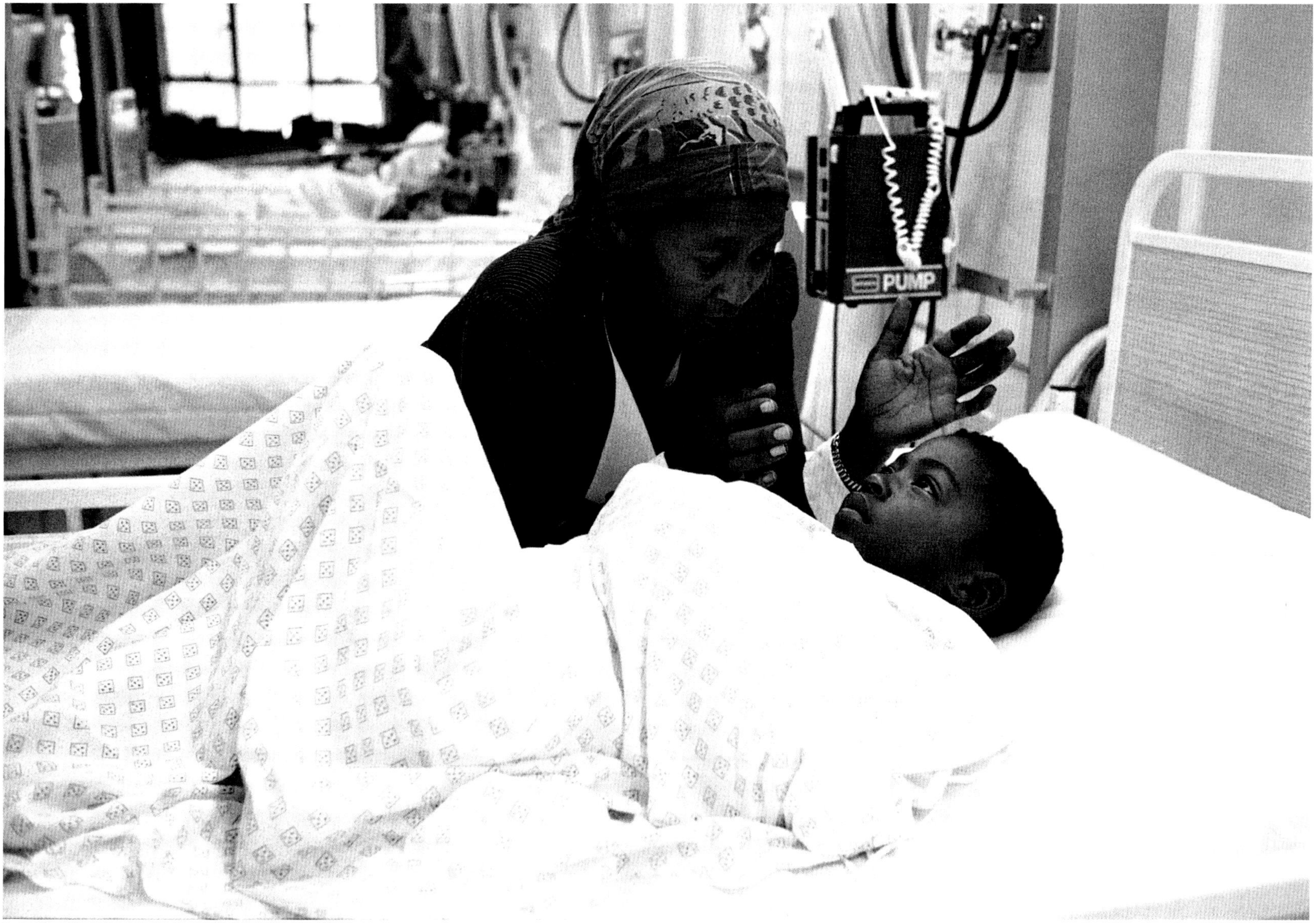

Red Cross Children's Hospital

Trauma Unit, reception area

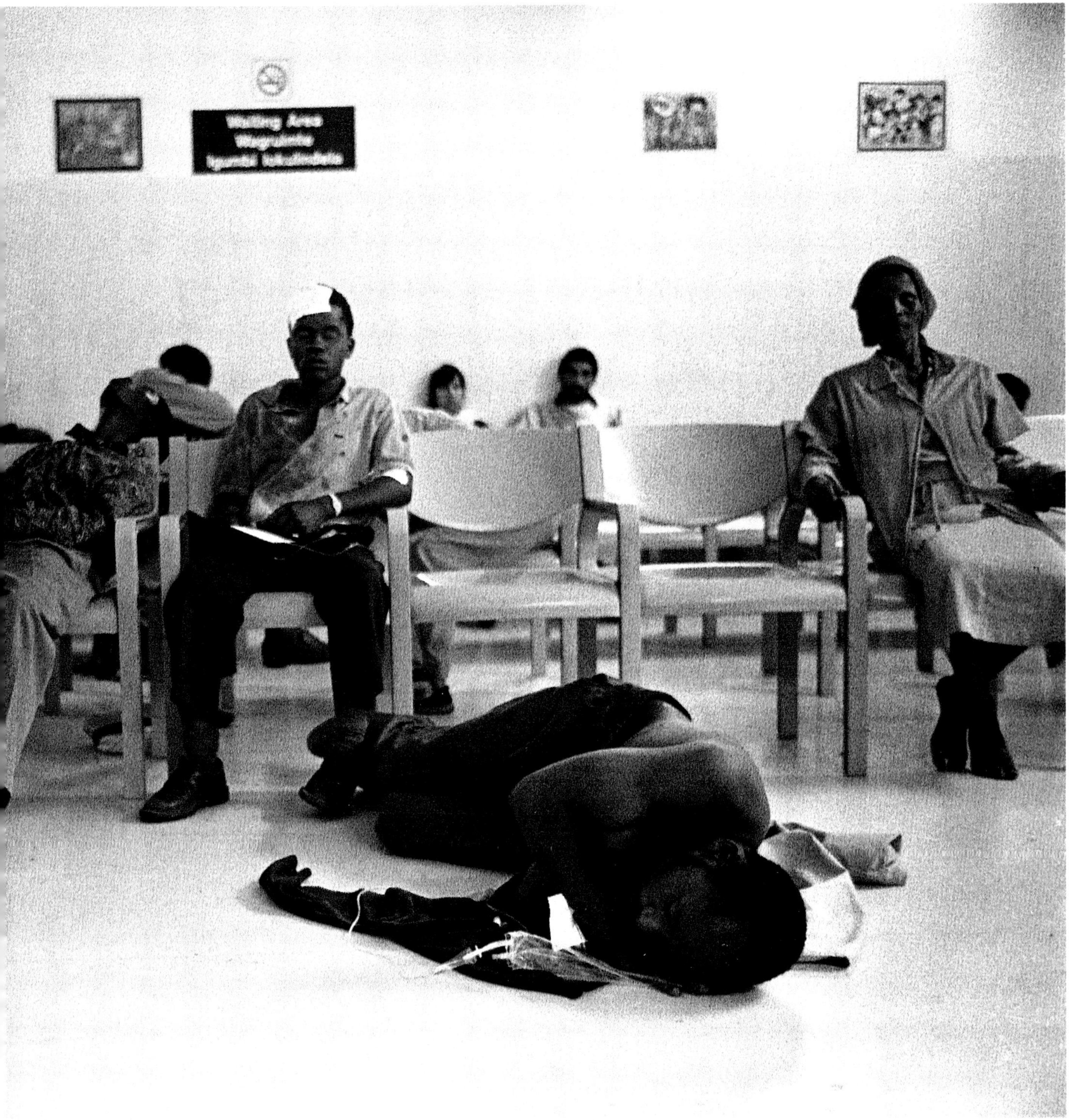

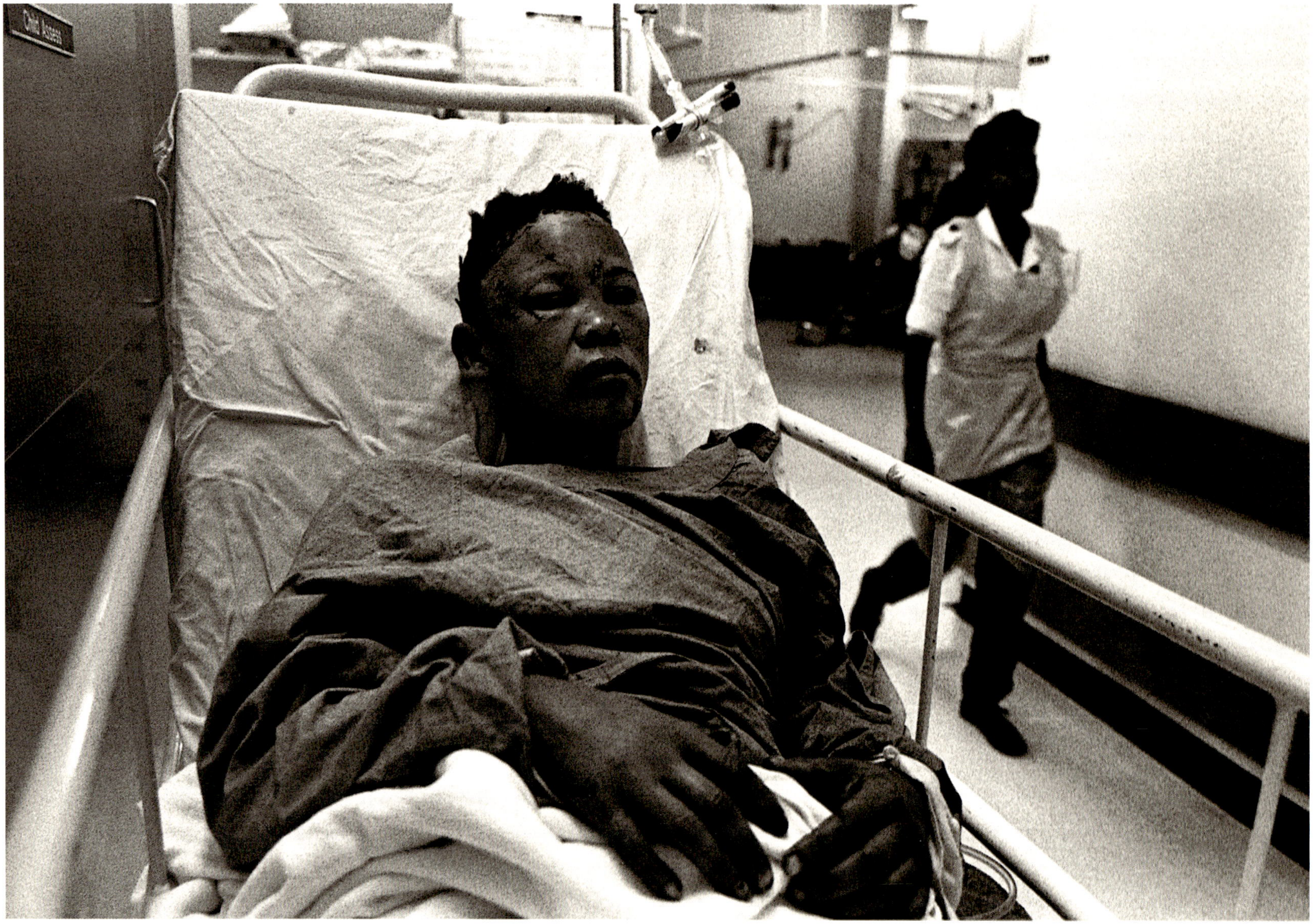

Trauma Unit, waiting to be treated

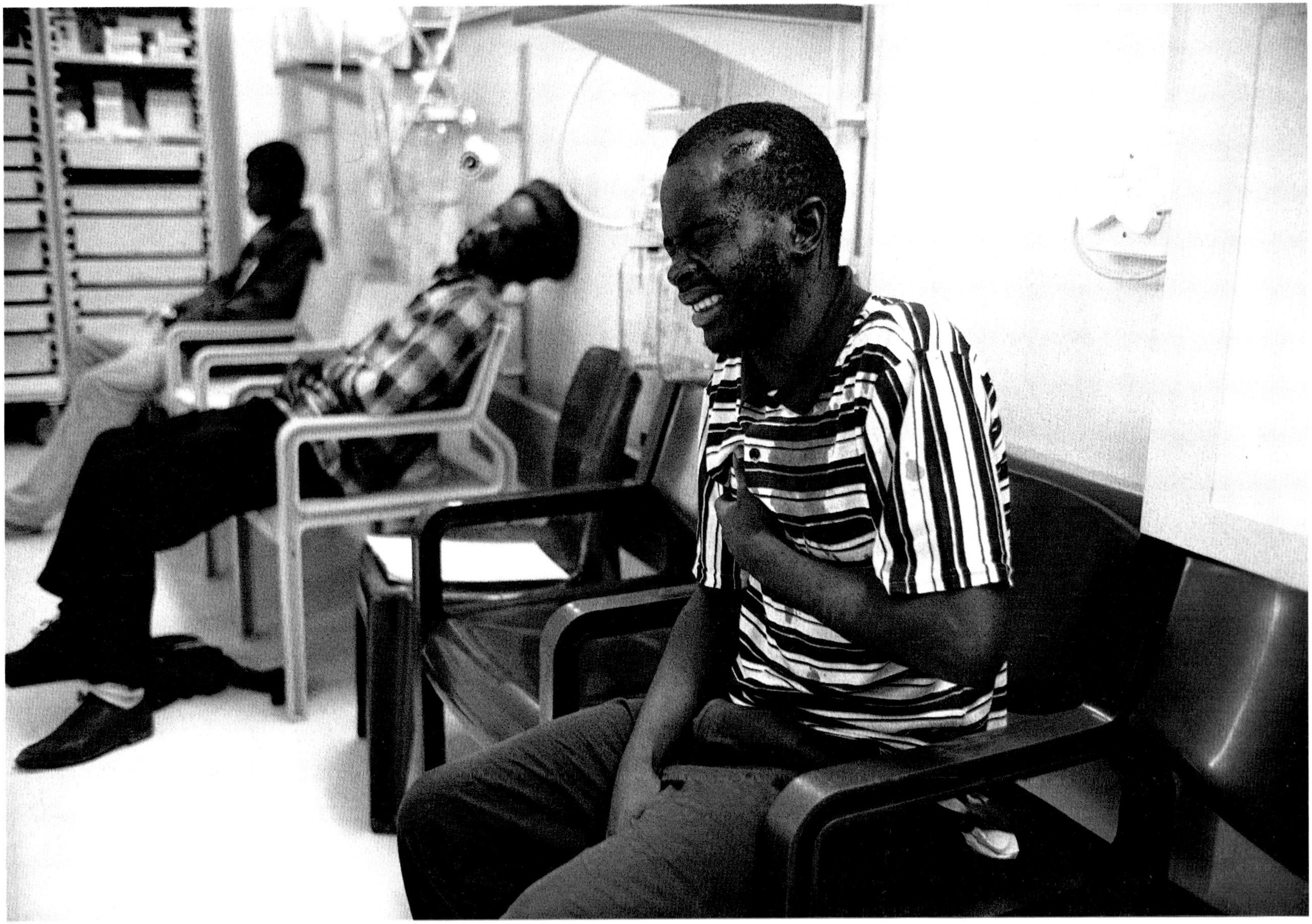

Trauma Unit, waiting to be treated

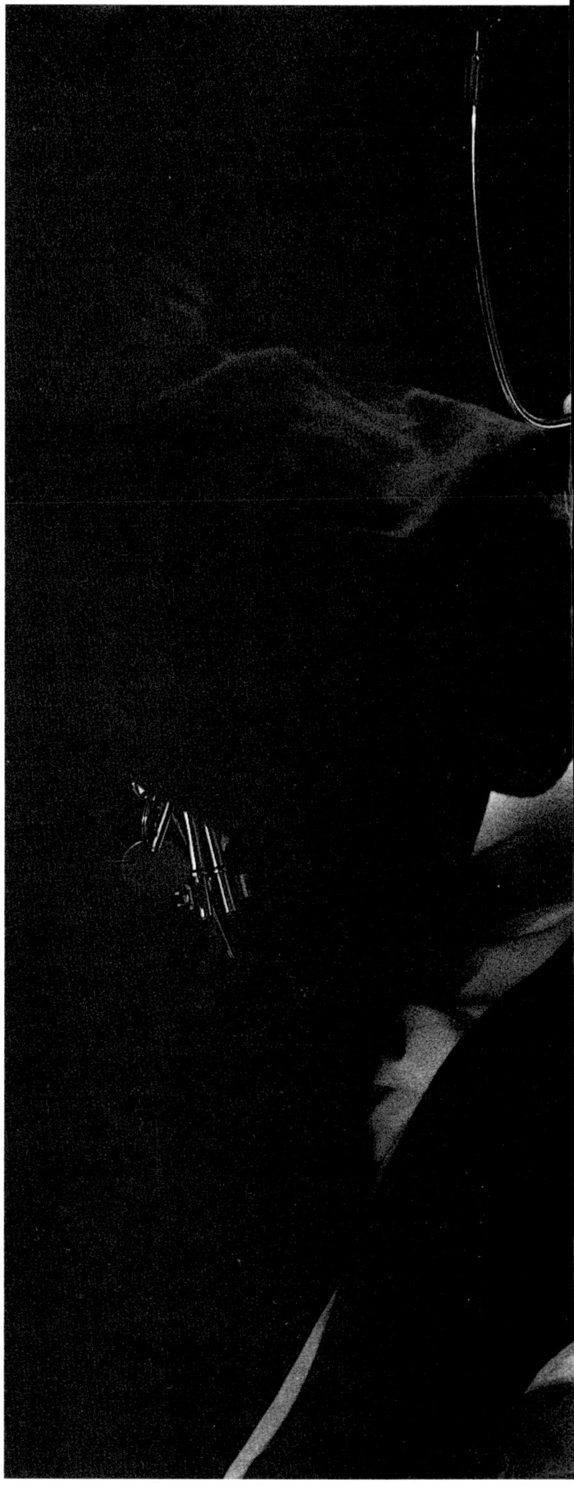

Groote Schuur Hospital, Trauma Unit (non-medical emergency, ie. violent assault). The operating theatre in the early hours of (any) Sunday morning

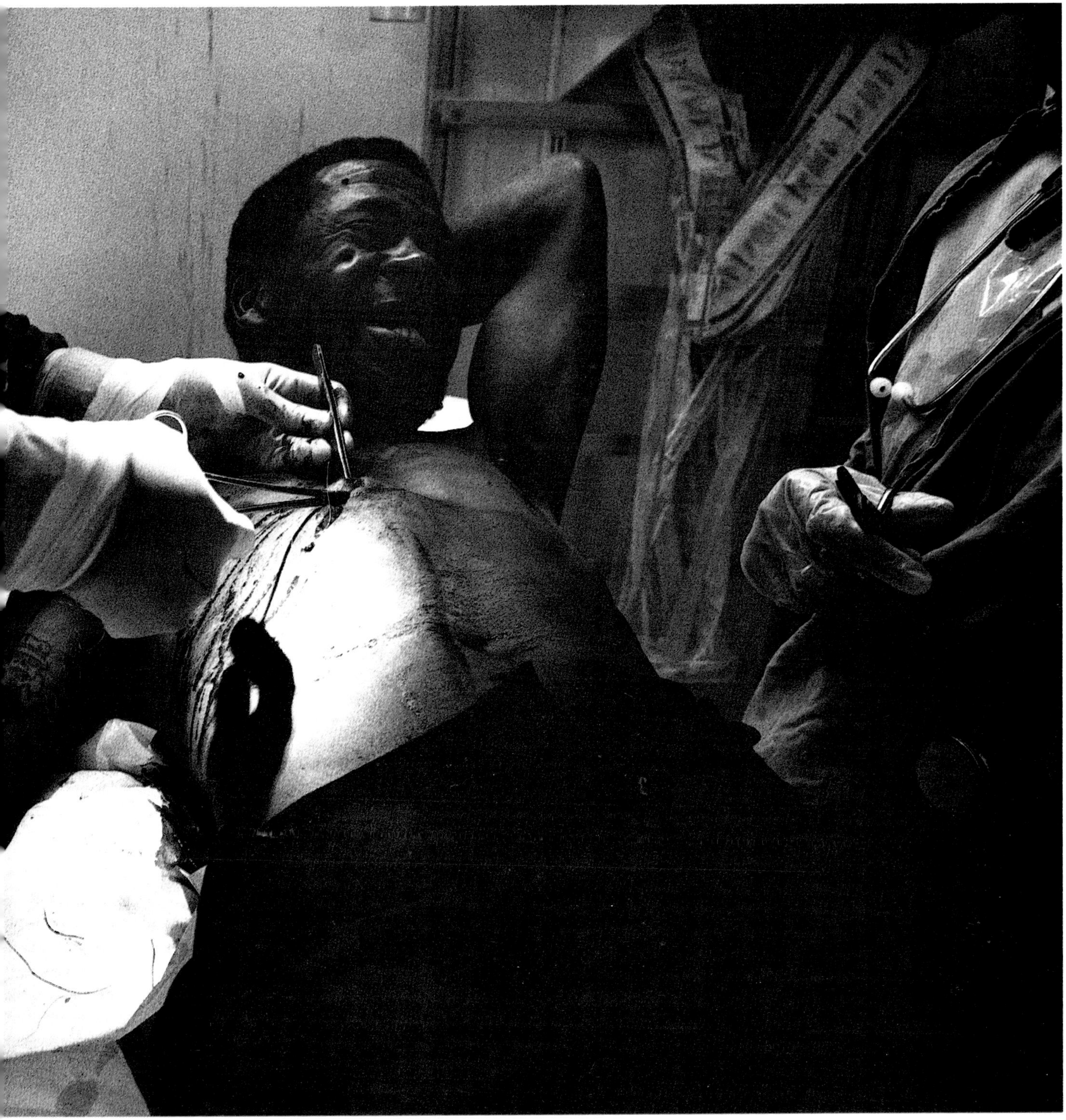

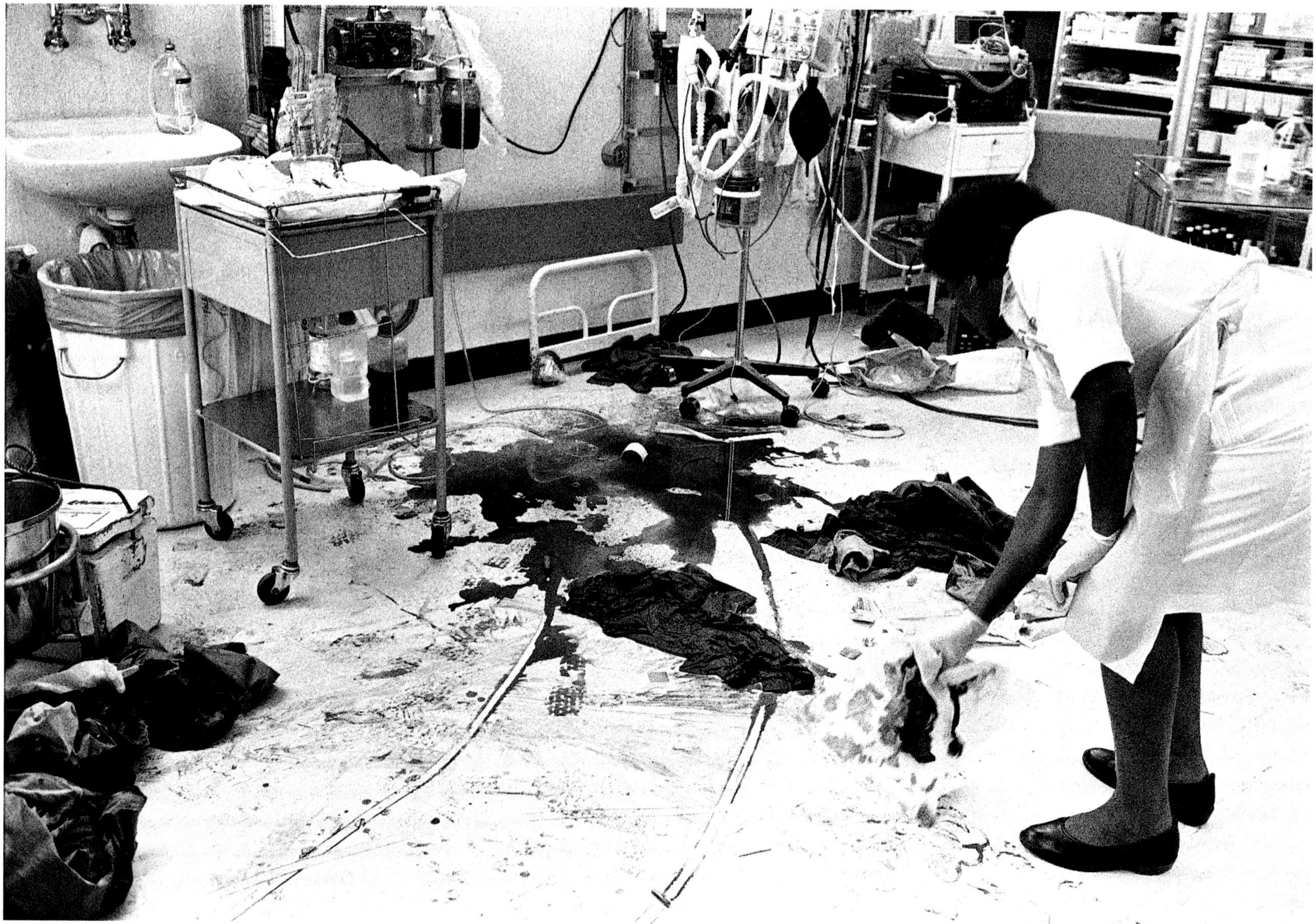

Aftermath, operating theatre of Trauma Unit

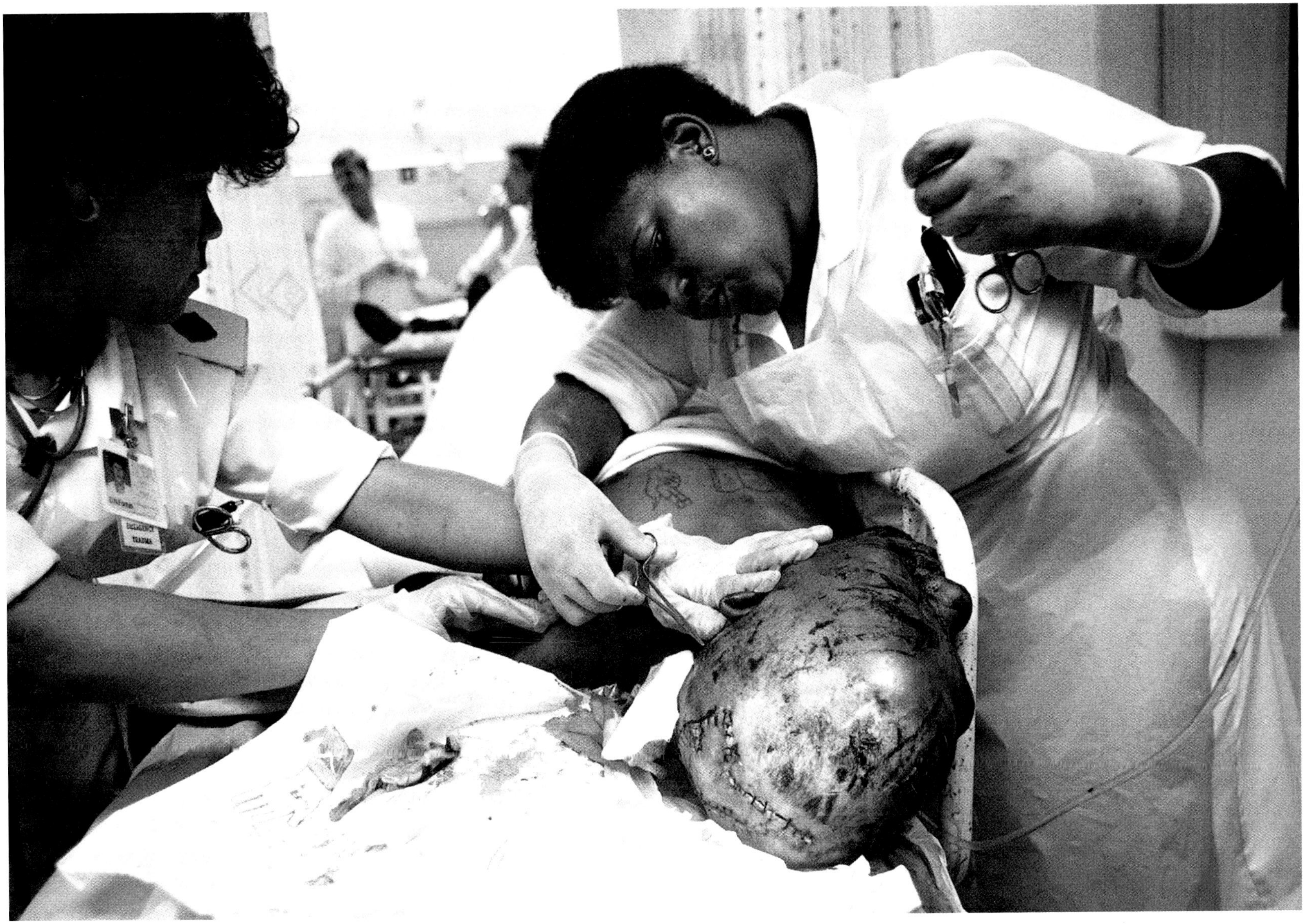

Groote Schuur Hospital, Trauma Unit

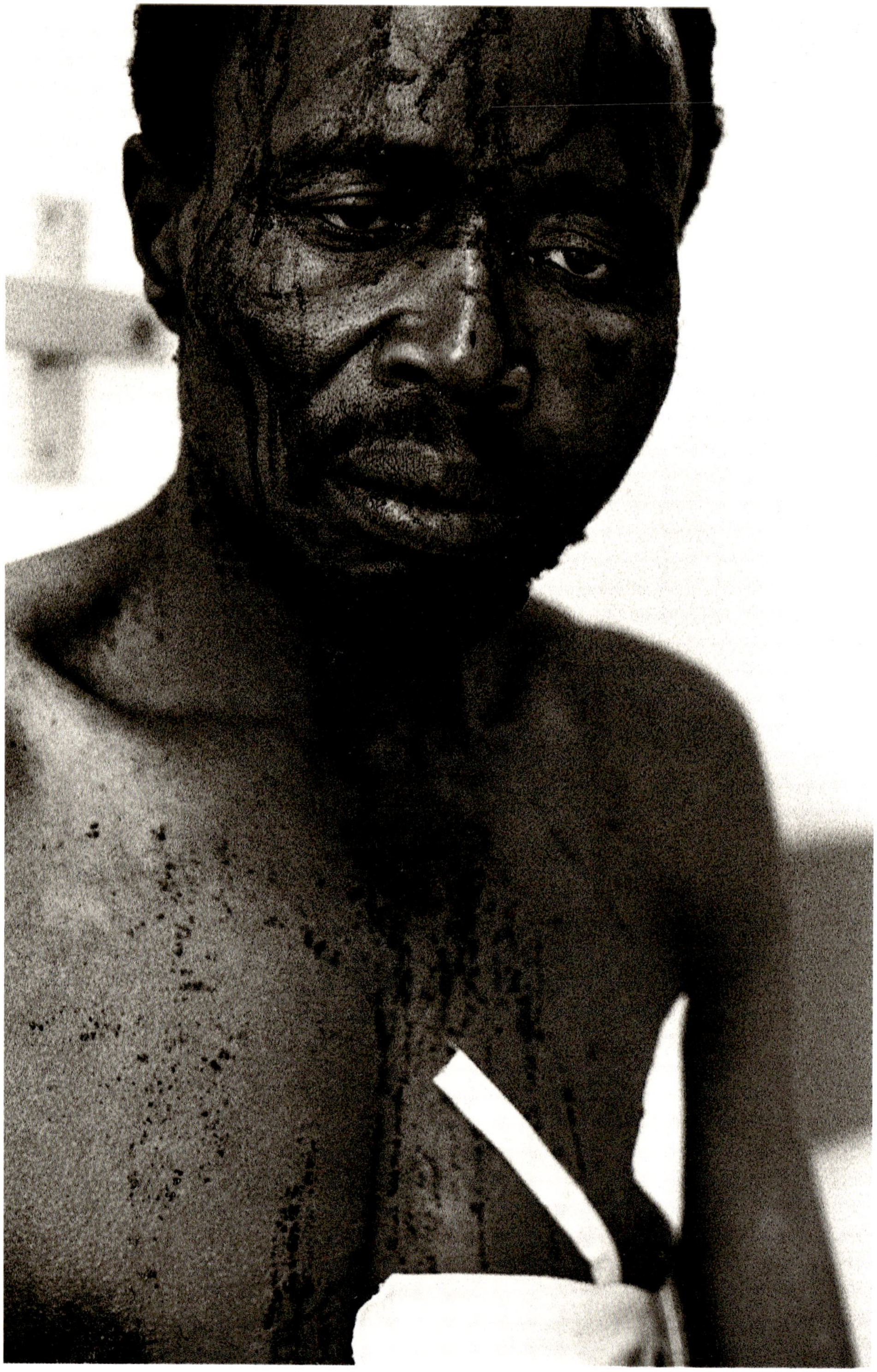

Trauma Unit

Trauma Unit

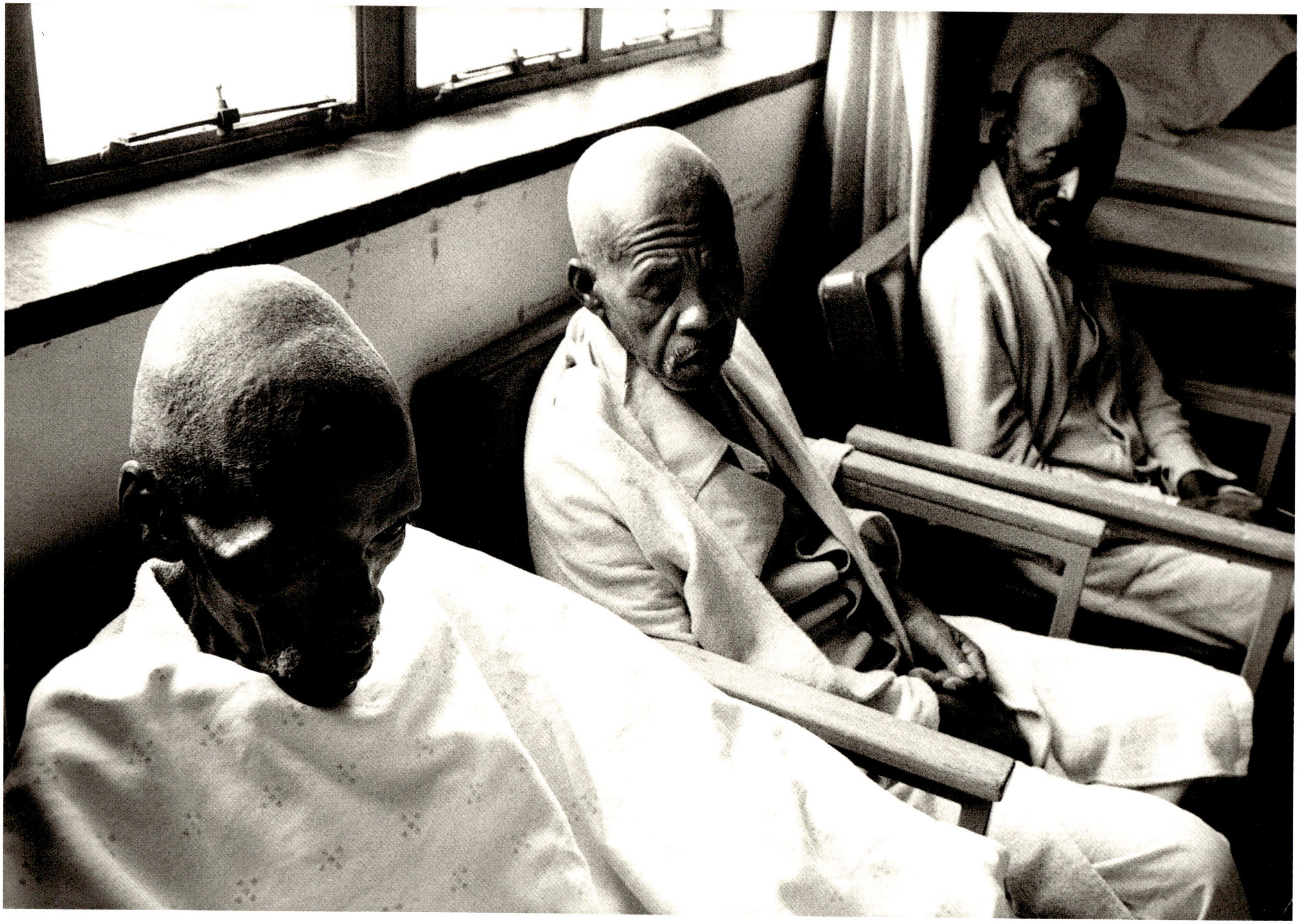

Tuberculosis Clinic

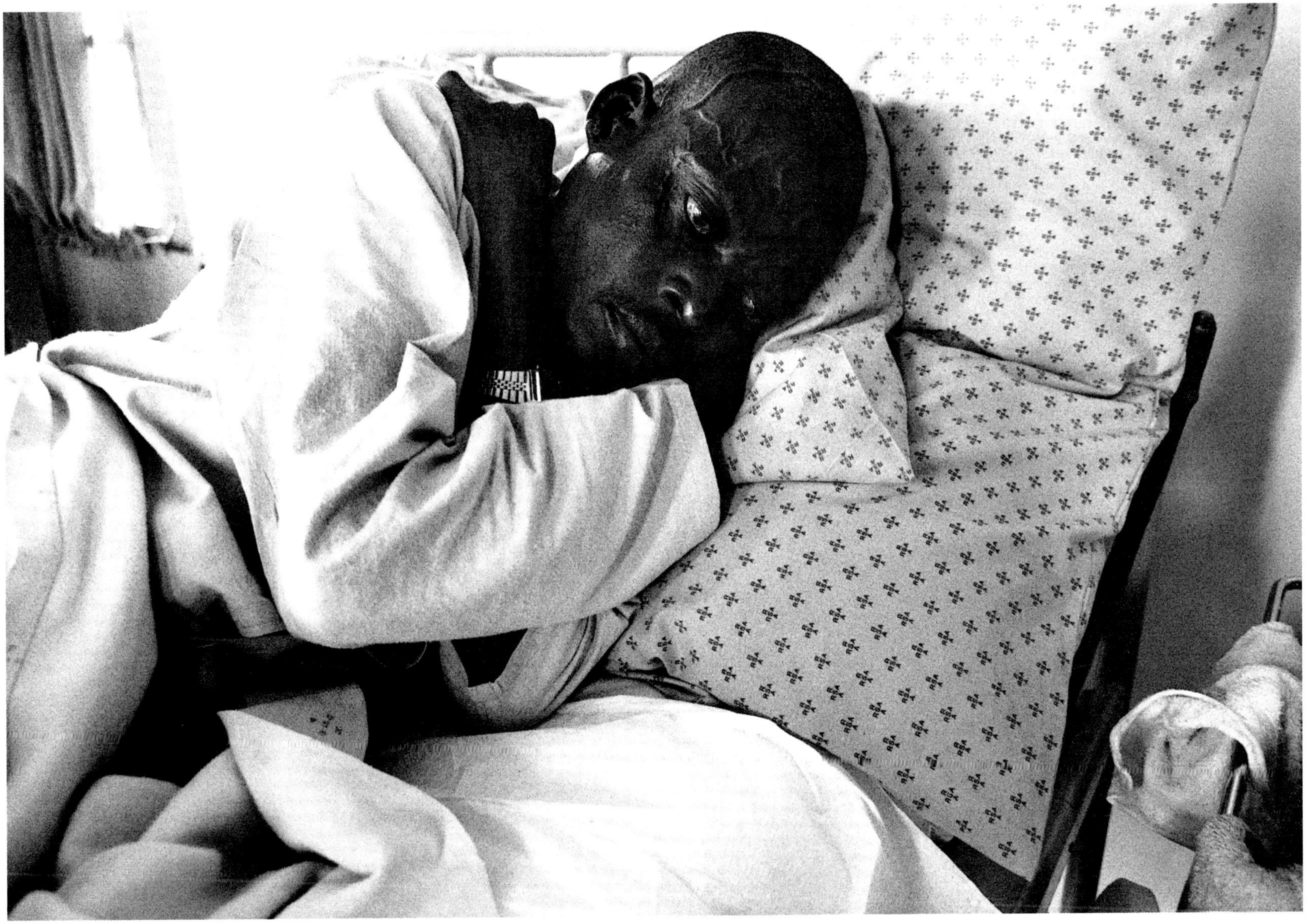

Tuberculosis Clinic

Waiting room, Khayelitsha Day Hospital

Early morning, Outpatients' section of the Khayelitsha Day Hospital

First published in 1994 by
Cornerhouse Publications
70 Oxford Street
Manchester M1 5NH England
061 237 9662

ISBN 0 948797 28 2

Prints: Joakim Blockstrom
Design and Artwork Production: Cornerhouse, UK
Reprographics: Beith Process, Johannesburg
Print: Creda Press, Cape Town

A complete catalogue of Cornerhouse Publications
is available on request